# Estuary Animals

Biome Beasts

Lisa Colozza Cocca

Rourke Educational Media
A Division of Carson Dellosa Education
rourkeeducationalmedia.com

# ROURKE'S SCHOOL to HOME CONNECTIONS
# BEFORE AND DURING READING ACTIVITIES

## Before Reading: *Building Background Knowledge and Vocabulary*

Building background knowledge can help children process new information and build upon what they already know. Before reading a book, it is important to tap into what children already know about the topic. This will help them develop their vocabulary and increase their reading comprehension.

## Questions and Activities to Build Background Knowledge:

1. Look at the front cover of the book and read the title. What do you think this book will be about?
2. What do you already know about this topic?
3. Take a book walk and skim the pages. Look at the table of contents, photographs, captions, and bold words. Did these text features give you any information or predictions about what you will read in this book?

## Vocabulary: *Vocabulary Is Key to Reading Comprehension*

Use the following directions to prompt a conversation about each word.
- Read the vocabulary words.
- What comes to mind when you see each word?
- What do you think each word means?

**Vocabulary Words:**
- adaptations
- bacteria
- brackish
- hydrodynamic
- migrate
- mudflats
- prey
- recedes
- teeming
- vertically

## During Reading: *Reading for Meaning and Understanding*

To achieve deep comprehension of a book, children are encouraged to use close reading strategies. During reading, it is important to have children stop and make connections. These connections result in deeper analysis and understanding of a book.

 Close Reading a Text

During reading, have children stop and talk about the following:
- Any confusing parts
- Any unknown words
- Text to text, text to self, text to world connections
- The main idea in each chapter or heading

Encourage children to use context clues to determine the meaning of any unknown words. These strategies will help children learn to analyze the text more thoroughly as they read.

When you are finished reading this book, turn to the next-to-last page for **Text-Dependent Questions** and an **Extension Activity**.

# Table of Contents

Biomes...................................................4
The Shoreline..........................................6
Mud Life..................................................9
Open Water............................................15
Land and Water......................................21
Activity: Mudflat Experiment....................29
Glossary................................................30
Index.....................................................31
Text-Dependent Questions.....................31
Extension Activity..................................31
About the Author...................................32

# Biomes

An estuary biome forms where rivers or streams meet an ocean. It is a mix of fresh water and salt water with water constantly moving in and out. It includes the **mudflats** that form when the water **recedes**.

Plants here must tolerate the **brackish** water. Pickleweed, saltgrass, eelgrass, and saltbush all do well here.

### Did You Know?
Estuaries come in all shapes and sizes. Some bays, harbors, inlets, lagoons, and sounds are estuaries.

# The Shoreline

Many birds live here. The great egret, a wading bird, tolerates fresh water and salt water. The great egret feeds at dusk. It wades into the shallow water and stands still. When **prey** comes near, it quickly thrusts its strong beak into the water and eats its catch whole. The great egret eats fish, frogs, snakes, salamanders, and crustaceans.

The western gull lives in estuaries along the Pacific coast of North America. These large gray and white birds have pink legs. They feed mainly on the surface of the water, but will do some shallow diving.

Western gulls are not picky eaters. These birds eat fish, clams, crabs, and chicks. They even steal food from seals and other birds.

### Did You Know?
The Canada goose lives in the estuary part-time. In mid-summer, it loses all its flight feathers. The estuary gives them a safe place to stay until their feathers grow back.

# Mud Life

The muddy bottom of the estuary and the mudflats are **teeming** with life. The sunflower star, one of the largest sea stars, lives here. It grows up to 39 inches (one meter) from arm tip to arm tip.

The sunflower star is born with five arms, but can grow 24 arms by adulthood. Its soft, spongy skin comes in many colors. If you pick one up, it droops. It depends on water pressure to keep its shape.

Most sea stars have a one-piece skeleton, but the sunflower star skeleton has several parts. This allows the sea star to open its mouth wide and expand its body to take in large prey. The star can swallow a sea urchin whole and then spit out its shell!

### Did You Know?

Estuary seahorses don't swim. They ride the water current. When in danger, the seahorse anchors its tail in the mud and hides among the plants growing there.

Newly hatched fiddler crabs swim **vertically** through the estuary to the open sea, where they have fewer predators. As adults, they return to the mudflats. They live in burrows dug at an angle to the surface. They eat bits of dead plants and animals in the water that washes over the flats.

### Did You Know?
One claw on male fiddler crabs is much larger than the other, which is used for eating. The smaller claw moves food into the mouth, which looks like a bow sweeping across a fiddle.

Newly hatched clams float in water, but move into the mudflats when still young. They dig burrows in the mud, which become their homes for life.

*The American oystercatcher looks for mollusks to eat, including clams and oysters.*

Clams eat when water washes over the mudflat. Their hinged, two-piece shells open and their foot-like necks stick out. Water fills the shells. The clams remove **bacteria** and other food matter and push out the leftover water.

### Did You Know?

Mud shrimp burrow deep into the mudflats to stay safe during low tides. One of their two sets of antennae is very long, though, and can sometimes be seen sticking out of the burrow.

# Open Water

Many fish and mammals live in estuary waters. One mammal is the manatee. It is often called a sea cow, but it is actually related to elephants. It uses its front flippers to steer as its flat tail pushes it forward through the water. It can stay underwater for about 15 minutes resting or for about four minutes swimming, but it must surface to breathe.

The manatee has small eyes and no outer ears, but has excellent sight and hearing. Its huge lips pick and move grasses into its mouth. Its teeth are only used for chewing and are replaced when worn down. The manatee eats one-tenth of its weight in sea grasses every day.

The estuary stingray lives in Australia. Its good, low-light vision is one of many **adaptations** that make it an excellent hunter. Its eyes are on top of its head, so the stingray can see prey swimming above. Its touch and smell senses are on its underside. They sense prey on the estuary floor. This stingray even senses water streaming out of clams and oysters.

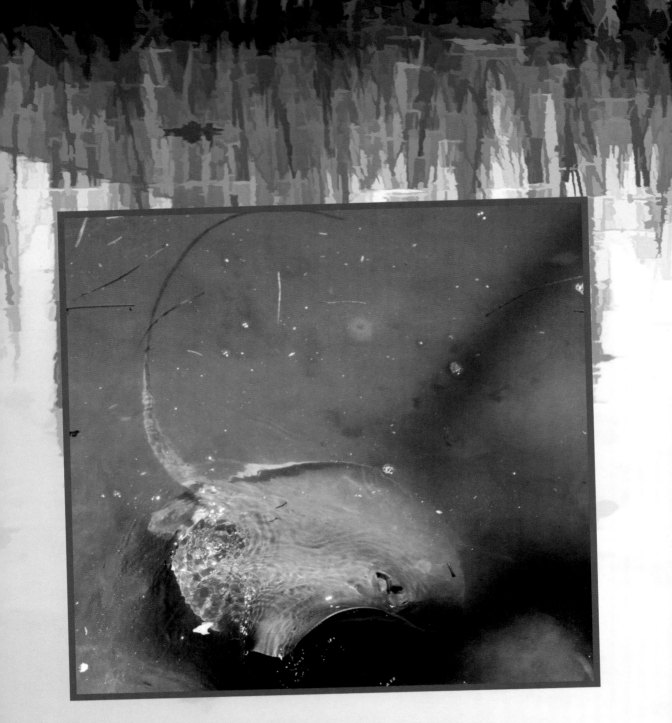

Estuary stingrays can't float. If they are not swimming or gliding, they sink to the bottom. This is an advantage when avoiding predators. It allows them to hide on the muddy bottom for long periods of time. Special openings behind their eyes and gills allow them to breathe while hiding.

Chinook salmon live part-time in the estuary. Adults spend most of their lives in the open ocean. When it is time to spawn, or deposit eggs, they **migrate** through the estuary to a freshwater stream. They spend time in the estuary to adjust from salt water to fresh water.

## Did You Know?

Estuaries are dangerous places for salmon. Large birds, seals, or bears gather there to catch the fish. The salmon's best defense is its oversized tail, which allows it to swim in fast bursts.

A few months after hatching, the young salmon migrate to the estuary. They stay there for several weeks or months until they are ready for the open ocean.

# Land and Water

Some estuary animals live both in water and on land. River otters run on land and swim in water. They have waterproof fur, webbed feet for paddling, and nostrils and ears that close in water.

River otters dig burrows with many tunnels on the shore. One tunnel opens underwater. This helps them escape land predators.

River otters hunt at night. Their long whiskers sense prey in the dark. Their clawed feet grip a slippery catch of fish, frogs, or turtles.

Harbor seals also divide their time between land and water. A harbor seal's small flippers are excellent for swimming, but not for walking. It flops around on its belly on land. In the water, it uses its front flippers to steer while its back flippers move it forward.

Harbor seals swim on their bellies and their backs. They can dive 1,500 feet (457 meters) underwater to find food. They eat most kinds of fish and sea creatures. The harbor seal finds prey by moving its whiskers back and forth. It can stay underwater for 40 minutes.

Mudskippers are walking fish. They breathe through gills when in the water. Before going on land, the mudskipper fills sacs around its gills with water bubbles. It fills its mouth with water.

On land, it breathes through its skin and the lining of its mouth. The bubbles in the sacs provide extra air as needed. The mudskipper can stay on land up to three days at a time.

### Did You Know?

Mudskippers hunt on mudflats using their **hydrodynamic** tongues. They spot a worm or crab and spit water over it. Then, the mudskipper quickly sucks up the water and the surprised prey.

Estuary biomes are filled with living things that have adapted to the constantly moving brackish water. They provide food and shelter to a wide variety of animals.

pelican

# ACTIVITY: Mudflat Experiment

The constant movement of water in the estuary means the mudflats undergo many changes. Complete this experiment to learn more about how those changes can affect the creatures living in the mudflats.

## Supplies

- deep, oblong baking pan
- sand
- pitcher of water
- six pebbles
- toothpick
- drinking straw

## Directions

1. Add some sand to one end of the pan. Do not go above half the pan height.
2. Slowly pour water over the sand until it cannot absorb any more water.
3. Use the toothpick to push three of the pebbles into the sand at different depths.
4. Use the drinking straw to make three holes, or tunnels, in the sand. Drop a pebble into each hole.
5. Pour water into the empty half of the pan, staying slightly below the height of the sand.
6. Tip the pan so water flows over the sand and hold for one minute.
7. Lay the pan flat so water returns to the empty side, tilting it slightly if needed.
8. Repeat steps 5 to 7 three times.

What happens to the hidden pebbles? Are some exposed? How did the tunnel width affect the results? How did the tunnel depth affect the results?

# Glossary

**adaptations** (ad-ap-TAY-shuhns): changes living things go through that allow them to better fit into their surroundings

**bacteria** (bak-TEER-ee-uh): single-celled living things that can cause good or harm

**brackish** (BRA-kish): somewhat salty, like the mix of fresh water and seawater found in estuaries

**hydrodynamic** (hye-droh-dye-NAM-ik): powered by the forces in or the movement of liquid

**migrate** (MYE-grate): move from one area to another

**mudflats** (MUHD-flats): flat lands that are covered by and then left bare by ocean tides

**prey** (pray): an animal that is hunted by another animal for food

**recedes** (ri-SEEDS): moves back

**teeming** (TEEM-ing): busy with the movement of many animals or people

**vertically** (VUR-ti-kuh-lee): straight up and down

# Index

bird(s) 6, 7, 8, 20
crab(s) 8, 12, 27
manatee(s) 15, 16
mudskipper(s) 26, 27
river otters 21, 22, 23

salmon 19, 20
seahorse(s) 10
seal(s) 8, 20, 24, 25
stingray(s) 17, 18
sunflower star 9, 10

# Text-Dependent Questions

1. Why are western gulls described as not being "picky eaters"?
2. How is a sunflower star different from most sea stars?
3. Describe the motion of the manatee as it swims forward in the sea.
4. Why is not being able to float an advantage for estuary stingrays?
5. Why are estuaries dangerous places for salmon?

# Extension Activity

Learn about an estuary near you. What animals live there? Draw a picture of the estuary. Add labels to show what lives where. Add an interesting fact about each animal.

# About the Author

Lisa Colozza Cocca has enjoyed reading and learning new things for as long as she can remember. She lives in New Jersey by the coast. She still gets excited every time she sees a fiddler crab making its way home. You can learn more about Lisa and her work at www.lisacolozzacocca.com.

© 2020 Rourke Educational Media

All rights reserved. No part of this book may be reproduced or utilized in any form or by any means, electronic or mechanical including photocopying, recording, or by any information storage and retrieval system without permission in writing from the publisher.

www.rourkeeducationalmedia.com

PHOTO CREDITS: Cover, page 1: ©33Karen33, ©Kevin Dyer, ©Gomez David; Graphics: ©KenSchulze; page 4-5: ©JavierGillooo; page 5: ©PhilAugustavo; page 6: ©tahir abbas; page 7: ©schmez; page 8: ©Pascale Gueret; page 8b: ©Spondylolithesis; page 9: ©NOAA; page 10: ©Fiona Ayerst; pages 10-11: ©Derek Holzapfel2012; page 13: ©BrianLasenby; page 14: ©onepony; page 14 (b): ©Wiki; page 15: ©Amanda Cotton; page 16: ©tobiasfrei; page 17: ©Jude Black; page 18: ©Wiki; page 19: ©mlharing; page 20: ©Supercaliphotolistic; page 21: ©milehightraveler; page 22: ©Brent Paull; page 23: ©Alec Taylor; page 24-25: ©RelaxFoto.de; page 26-27: ©PatrickGijsbers; page 27: ©j_pich; page 28: ©FlorinEne; page 28-32: ©HTakemoto

Edited by: Kim Thompson
Cover design by: Kathy Walsh
Interior design by: Rhea Magaro-Wallace

**Library of Congress PCN Data**

Estuary Animals / Lisa Colozza Cocca
(Biome Beasts)
 ISBN 978-1-73161-442-1 (hard cover)
 ISBN 978-1-73161-237-3 (soft cover)
 ISBN 978-1-73161-547-3 (e-Book)
 ISBN 978-1-73161-652-4 (ePub)
Library of Congress Control Number: 2019932142

Rourke Educational Media
Printed in the United States of America,
North Mankato, Minnesota